ISBN 978-0-260-61623-4
PIBN 11119134

English
Français
Deutsche
Italiano
Español
Português

www.forgottenbooks.com

Mythology Photography **Fiction**
Fishing Christianity **Art** Cooking
Essays Buddhism Freemasonry
Medicine **Biology** Music **Ancient
Egypt** Evolution Carpentry Physics
Dance Geology **Mathematics** Fitness
Shakespeare **Folklore** Yoga Marketing
Confidence Immortality Biographies
Poetry **Psychology** Witchcraft
Electronics Chemistry History **Law**
Accounting **Philosophy** Anthropology
Alchemy Drama Quantum Mechanics
Atheism Sexual Health **Ancient History**
Entrepreneurship Languages Sport
Paleontology Needlework Islam
Metaphysics Investment Archaeology
Parenting Statistics Criminology
Motivational

Relationships Between Preslaughter and Postslaughter Evaluations of Beef Cattle

Circular No. 945   MAY 1954

UNITED STATES DEPARTMENT OF AGRICULTURE

CONTENTS

For sale by the Superintendent of Documents, U. S. Government Printing Office
Washington 25, D. C. - Price 15 cents

Relationships Between Preslaughter and Postslaughter Evaluations of Beef Cattle[1]

By R. R. WOODWARD, J. R. QUESENBERRY, R. T. CLARK, C. E. SHELBY, and O. G. HANKINS[2]

ANIMAL AND POULTRY HUSBANDRY RESEARCH BRANCH, AGRICULTURAL RESEARCH SERVICE

INTRODUCTION

Experimental studies on production records for beef cattle first were outlined approximately 20 years ago. At that time several plans were submitted for keeping records of performance to supplement selection based purely on conformation and visual appraisal. Although the original suggestions as to the most important objective measurements for determining productivity have been modified and revised, much of the experimental work relating to the problem was stimulated.

Experiments were designed to evaluate measurements of production in beef cattle by records of performance. Such factors as birth weight, weaning weight and grade, rate and efficiency of gain on feed, weight at end of feeding period, weight at 18 months, and slaughter grade have been found of varying importance as production indices.

Carcass studies have been included in some record-of-performance experiments. However, considerable carcass appraisal work has been carried out without preslaughter records. Similarly, information on growth rates often has been collected without a corresponding carcass analysis. In general there have been more data collected and analyzed for the preslaughter period than for the postslaughter period. The record-of-performance work has proceeded without definite knowledge of the effect of selection for rapid gains on the type of carcass

[1] These data represent part of the work done by the senior author in fulfilling the requirement for degree of doctor of philosophy at the University of Minnesota.

[2] The authors wish to acknowledge the assistance of A. L. Baker and Bradford Knapp, Jr., of the Animal and Poultry Husbandry Research Branch, Agricultural Research Service; of F. S. Willson of the Montana Agricultural Experiment Station; and of L. M. Winters of the University of Minnesota.

O. G. Hankins died November 13, 1953.

produced, regardless of the possibility that such selection may be antagonistic to the development of improved carcasses. Preslaughter data are both easier and less expensive to obtain. Furthermore, selection for rate and efficiency of gain has been effective and a number of beef-cattle improvement projects have been initiated independent of a progeny test.

This study was designed to explore the relationships between preslaughter and postslaughter evaluations of a group of Hereford steers. Preslaughter evaluations were based principally upon growth records while the calf was suckling its dam as well as during an extended feeding period. Postslaughter data were comprised of both subjective and objective carcass evaluations of the same steers.

MATERIALS AND METHODS

The data reported in this study were collected at the United States Range Livestock Experiment Station at Miles City, Mont. The period covered was from 1942 to 1951, inclusive. The animal breeding experiments at this station are conducted cooperatively by the Animal and Poultry Husbandry Research Branch, Agricultural Research Service of the United States Department of Agriculture and the Montana Agricultural Experiment Station. The cattle included in this report were all unregistered Herefords.

The data were taken from records of performance on steer progeny of bulls from inbred Hereford lines maintained by the station. Records were kept on the steers from birth and included postslaughter evaluations of the carcasses. The inbred Hereford lines were started by purchases from registered herds in a number of the Western States. An effort was made to obtain samples of herds of varying types to allow a more accurate appraisal of variations existing within the Hereford breed. Some of the bulls were from newly established lines and others from lines that have been closed as long as 17 years. Bulls from certain lines were selected as herd sires on the results of these progeny tests. The final selection was primarily dependent upon an index composed of weaning weight, rate of gain, and visual appraisal at the end of a time-constant feeding period.

Management of the Herd

With few exceptions the calves included in each progeny test were born within a period of 6 weeks. Calving started approximately on March 25 each year. All calves were weighed within 24 hours after birth. Male calves were castrated in May or the first week of June.

Each member of a cow herd was selected at random except to balance each group for age. Approximately 30 cows were used in each test herd. Bulls were left in the herd for 45 days.

All of the cows were carried in common pastures except during the breeding period when they were separated into breeding pastures. Every effort was made to eliminate environmental variables such as pasture differences during the preweaning period.

All calves were weaned the same day and usually on the Monday closest to the 20th of October each year. After weaning weights were obtained each calf was individually scored by a committee of three members. Each sire group was appraised at the conclusion of the individual scoring.

2

Record-of-Performance Procedure

If available, eight steers to be tested were selected at random from each sire, as reported by Knapp et al. (*15*) [3] Obvious culls were eliminated. The first 10 days after weaning were used to accustom the steers to the grain and hay of the experimental ration before the regular trial started.

Initial weights were the average of three weighings taken 2 days prior to the trial and the first day of the trial.

The experimental ration consisted of:

6 parts corn or barley	1 part wheat bran
3 parts dried beet pulp	alfalfa hay
1 part linseed oil meal	

A commercial vitamin-A supplement was included in the concentrate portion of the ration, and a mixture of salt and bonemeal (2 parts of salt to 1 part of bonemeal) was fed free choice.

Table 1 lists the number of steers in the trial, the grain used in the rations, and the duration of trial.

TABLE 1.—*Steers in feeding trials: Number, kind of grain in rations, and duration of trials, 1941–51*

Year	Steers tested	Grain	Duration of trial
	Number		*Days*
1941–42	62	Corn	252
1942–43	45	do	273
1943-44	46	do	259
1944–45	53	do	252
1945–46	63	do	252
1946–47	98	do	252
1947-48	85	Barley	252
1948–49	88	do	252
1949–50	50	Corn	252
1950–51	45	do	252
Total	635		

The steers were started on 2 or 3 pounds of concentrates and about 5 pounds of alfalfa. Fairly uniform amounts of feed were offered at the beginning of the trial, but as the trial progressed the amounts offered depended on individual appetites. At the conclusion of the trial they were consuming about five parts of concentrates to one part of roughage. In total pounds this approximated 17 to 18 pounds of concentrate to 3 to 4 pounds of roughage, with considerable deviation both below and above according to individual appetites. All steers were individually fed through 1946 and fed by sire groups thereafter. The concentrate portion of the ration was hand fed twice daily, and the roughage portion was hand fed once daily. Feed refused was weighed back daily.

The duration of the test was nine 28-day periods with the exceptions shown in table 1. These exceptions were adjusted to 252 days.

---

[3] Italic numbers in parentheses refer to Literature Cited, p. 18.

Individual weights were taken at the end of each 28-day period. At the conclusion of the trial the steers were weighed on each of the last 3 days of experimental feeding. The final weight represented the average of the three successive weights.

The steers were all marketed in South St. Paul, Minn. Shipping conditions were remarkably uniform considering the fact that little or no control was possible with the public carriers involved. The same sequence was followed each year for shipping and marketing. The trials were concluded on a Wednesday, the cattle were loaded for shipment within 24 hours after the last weight was taken, and they reached St. Paul on the following Saturday or Sunday. On Monday they were individually graded by a committee of three members. One committee member graded every year of the experiment and several other members graded at least 5 of the 10 years. Grades were made according to P. M. A.-B. A. I.-A. R. A. chart No. 101.

The steers were sold by sire groups which, in effect, resulted in a packer-buyer appraisal of the progeny of each bull. Not all steers in a sire group necessarily sold for the same price per pound. Sale weights were taken in the yards the afternoon the cattle were sold. The steers were taken from the scales to the packinghouse immediately after being weighed and were killed the following day except for 1 year when they were sold to a small plant which killed the next 2 successive days. On the killing floor, the eartags were transferred to the briskets of the carcasses to maintain the identities of the steers. Care was necessary to avoid loss of the tag during the washing and shrouding of the carcass. As an additional precaution the sequence in which the steers were killed was recorded. Hot carcass weights were recorded on the killing floor.

The carcasses were held in coolers for 48 hours before carcass data were taken. On the morning of the second day, the right sides of the carcasses were divided (ribbed) between the 12th and 13th rib.

The following carcass data were recorded:

CARCASS GRADES.—The grading was done by a committee of three members. As in the slaughter grading committee, one member served each year of the experiment. Most of the other members graded at least 2 of the 10 years. Grades were entered on A. M. A.-B. A. I. chart No. 102. The final slaughter and carcass scores given each steer were an average of the scores of the three committee members.

COLOR OF LEAN.—Color readings were made from the eye muscle 30 minutes after the carcasses were ribbed, thus allowing the meat to brighten a uniform period of time. The number on the standard color disc most nearly matching the color of the lean of the eye muscle was recorded. Colors ranged from bright to dark red, with corresponding low to high numbers.

AREA OF EYE MUSCLE AND THICKNESS OF FAT OVER EYE MUSCLE.—Transparent, nonabsorbent tracing paper was laid on the posterior side of the 12th-rib cut, and tracings were made of the eye muscle and surrounding external fat. The area of the eye muscle was determined later by following the outline with a compensating polar planimeter. Thickness of fat was determined by averaging three thickness measurements made perpendicular to the longitudinal axis of the eye muscle at one-fourth, one-half, and three-fourths of the length of the axis.

4

BODY MEASUREMENTS.—Two measurements were made on each steer: Length of body and length of hind leg. Both measurements were made with a flexible steel tape graduated in millimeters. The length of body was measured from the anterior edge of the first thoracic vertebra to the anterior edge (lowest point) of the aitch bone. The length of hind leg was measured from the anterior edge (lowest point) of the aitch bone to the highest point of the hock joint (highest point on the tarsal bones in the hock joint).

CARCASS WEIGHTS.—Cold carcass weights were taken immediately after the preceding data were collected. The time averaged about 48 hours after the hot carcass weights were taken.

DATA AND DISCUSSION

The mean values of the various factors studied are presented by years and by lines within years in the appendix.

Listed below are the records considered in this study.

PRESLAUGHTER:

Birth weight (pounds).

Weaning weight (pounds).

Gain on test (pounds)—total gain during the feeding period.

Final weight (pounds)—obtained at the conclusion of the time-constant feeding period before cattle were shipped.

Efficiency (gain per 100 pounds of TDN—total digestible nutrients)—computed from total gain and total digestible nutrients consumed during the feeding period.

Sale weight (pounds)—obtained in South St. Paul after a 24- to 48-hour fill.

Slaughter grade (according to P. M. A.-B. A. I.-A. R. A. chart No. 101). Grades range from 1 to 42 and indicate high to low grades. Grades were given to individual steers just prior to slaughter.

POSTSLAUGHTER:

Carcass grade (according to A. M. A.-B. A. I. chart No. 102): Grades range from 1 to 42, indicating high to low.

Carcass weight (pounds) refers to weight of carcass after 48 hours in cooler.

Length of body (centimeters): Measured from anterior edge of first thoracic vertebra to anterior edge of aitch bone.

Length of leg (centimeters): Measured from anterior edge of aitch bone to highest point of hock joint.

Area of eye muscle (square inches): Area of longissimus dorsi between 12th and 13th ribs.

Thickness of external fat (millimeters): Average of three measurements of external fat covering eye muscle, taken at 25, 50, and 75 percent of the longitudinal axis of the eye muscle.

Color of lean (according to United States Department of Agriculture Color Standard A): Colors range from 1 to 10 indicating light to dark red color.

Dressing percentage (percent): Cold carcass weight divided by the final weight at Miles City.

Relationships Between Birth and Weaning Weights and Subsequent Growth and Carcass Evaluations

Birth Weight

Total, within line, within year, and within line and year correlations between birth weight and other factors are presented in table 2. A positive relationship was found between birth weight and weaning weight as well as between birth weight and rate of gain, and birth weight and final weight. These findings are in agreement with Woodward et al. (28), who stated that beef heifers larger at birth were also larger at weaning, and maintained their weight advantage until and after maturity.

TABLE 2.—*Correlations between birth weight and other factors studied* [1]

Factor	Simple correlations				Partial correlations [2]
	Total	Within lines	Within years	Within lines and years	Within lines and years
Weaning weight	0. 41*	0. 38*	0. 36*	0. 31*	--------
Final weight	. 39*	. 40*	. 48*	. 51*	--------
Gain on test	. 17*	. 23*	. 31*	. 43*	--------
Efficiency	—. 04	. 01	. 05	. 12*	--------
Slaughter grade	. 18*	. 14*	. 12*	. 07	--------
Carcass grade	—. 01	—. 04	—. 04	—. 07	--------
Area of eye muscle	. 28*	. 30*	. 26*	. 28*	0. 07
Thickness of fat	—. 12*	—. 13*	—. 07	—. 09	—. 24*
Length of body	. 38*	. 41*	. 40*	. 43*	. 06
Length of leg	. 43*	. 47*	. 45*	. 50*	. 23*

[1] Asterisk indicates significant correlation at 1-percent level.
[2] Computed with final weight held constant.

When line and year effects were removed there was a low but significant correlation between birth weight and efficiency.

Birth weight was not significantly correlated with slaughter or carcass grade when year and line effects were removed. A significant positive relationship existed between birth weight and area of eye muscle, but a negative relationship was found between birth weight and thickness of fat over the eye muscle.

Partial correlations were computed between birth weight and area of eye muscle and between birth weight and thickness of fat, holding final weight constant. There was still a positive correlation between birth weight and area of eye muscle, but it was just below the 5-percent level of significance. The intensity of the negative relationship between birth weight and thickness of external fat over the eye muscle (ribeye) was increased when final weight was held constant.

Correlations between birth weight and the body measurements taken on the carcass show that birth weight was not only indicative of future rate of growth but also of body proportions. In the population studied there was a higher correlation between birth weight and length of leg than between birth weight and length of body. When final weight was held constant the difference became even greater.

Weaning Weight

Under the conditions of this experiment there _was an extremely wide yearly variation in weaning weights, primarily because of the occasional years of drouth in the area in which the experiment was conducted. Under these occasional adverse conditions, it is doubtful if there is much expression of the genetic potential of the calf for differential growth prior to weaning. Fortunately during the 10 years of the experiment only 2 years were below the long-time mean for precipitation during the growing season.

In table 3 are given the correlations between weaning weight and other factors studied. Pontecorvo (25) reported a high correlation between the weights of cattle at 6, 12, and 18 months. However, Knapp et al. (14) stated that average daily gain or efficiency of gain in the feedlot could not be predicted from the previous rate of gain during the suckling period.

TABLE 3.—*Correlations between weaning weight and other factors studied* [1]

| Factor | Simple correlations | | | | Partial correlations [2] |
	Total	Within lines	Within years	Within lines and years	Within lines and years
Final weight	0. 55*	0. 61*	0. 60*	0. 68*	--------
Gain on test	. 03	. 12*	. 07	. 23*	--------
Efficiency	−. 45*	−. 42*	−. 36*	−. 29*	0. 40*
Slaughter grade	. 24*	. 22*	. 20*	. 17*	--------
Carcass grade	. 08	. 07	. 15*	. 16*	. 06
Area of eye muscle	. 34*	. 36*	. 30*	. 32*	. 26*
Thickness of fat	. 07	. 14*	. 12*	. 26*	. 25*

[1] Asterisk indicates significant correlation at 1-percent level.
[2] Computed with gain held constant.

In this study a significant positive correlation was obtained between weaning weight and rate of gain when line and year effects were removed. This would tend to confirm the heritability estimates for weaning weight proposed by Knapp and Nordskog (16) and Knapp and Clark (17).

The significant negative correlation between weaning weight and efficiency was to be expected. Because of the standard preweaning treatment, the growth of the calves from birth to weaning was largely dependent upon the milk-producing ability of their dams. Calves from the better milk producers carried more condition when they went into the feedlot.

Weaning weight was found to be related to slaughter grade, carcass grade, area of eye muscle, and thickness of fat over the eye. The correlation coefficients were relatively low for each factor. As the correlations were positive, there is no indication of antagonism between heavy weaning weights (probably a reliable indication of adequate milk production) and the ability to produce a desirable carcass. When gain was held constant the positive relationships still held, but the correlation with the carcass grade was reduced below the level of significance.

Relationships Between Performance in the Feedlot and Subsequent Carcass Evaluations

Gain on Test and Final Weight

Correlations between gain on test and carcass evaluations are shown in table 4. Correlations between final weight and carcass evaluations are shown in table 5.

TABLE 4.—*Correlations between gain on test and other factors studied* [1]

Factor	Simple correlations				Partial correlations [2]
	Total	Within lines	Within years	Within lines and years	Within lines and years
Efficiency	0. 23*	0. 22*	0. 42*	0. 47*	
Slaughter grade	. 30*	. 37*	. 36*	. 49*	
Carcass grade	. 35*	. 36*	. 40*	. 43*	
Area of eye muscle	. 29*	. 30*	. 33*	. 36*	0. 00
Thickness of fat	. 57*	. 50*	. 33*	. 07	−. 21*
Dressing percentage	. 30*	. 26*	. 11	−. 01	
Length of body	. 53*	. 57*	. 57*	. 64*	−. 03
Length of leg	. 44*	. 49*	. 49*	. 56*	−. 09

[1] Asterisk indicates significant correlation at 1-percent level.
[2] Computed with final weight held constant.

TABLE 5.—*Correlations between final weight and other factors studied* [1]

Factor	Total	Within lines	Within years	Within lines and years
Fficiency	−0. 05	−0. 05	0. 09	0. 12*
Slaughter grade	. 40*	. 43*	. 51*	. 57*
Carcass grade	. 34*	. 33*	. 40*	. 40*
Area of eye muscle	. 43*	. 43*	. 44*	. 44*
Thickness of fat	. 49*	. 45*	. 33*	. 22*
Dressing percentage	. 35*	. 36*	. 25*	. 25*
Length of body	. 78*	. 79*	. 78*	. 79*
Length of leg	. 70*	. 70*	. 70*	. 72*

[1] Asterisk indicates significant correlation at 1-percent level.

8

Winters and McMahon (*27*) reported a correlation coefficient of 0.7141 between rate and efficiency of gain. A similar association was observed by Stanley and McCall (*26*), although Knapp et al. (*14*) stated that daily gain and efficiency of gain were not highly related in a time-constant population.

In this study the correlation between gain and efficiency was much higher than between final weight and efficiency. Since final weight was essentially a composite of weaning weight and total gain, the difference was undoubtedly due to the negative relationship existing between weaning weight and efficiency.

There was a significant correlation between both gain and final weight with slaughter and carcass grades. Slaughter grades were more closely associated with final weight than with rate of gain. Carcass grades were also more closely associated with final weight than with rate of gain, but the difference was not so wide.

Hankins and Burk (*9*) found a correlation coefficient of 0.37 between rate of gain and carcass grade. This compares very closely with the total correlation coefficient of 0.35 obtained in this study even though the population studied by Hankins and Burk was composed of both sexes and of age groups from calves to 2-year-old steers. Hankins and Burk further concluded that rate of gain was correlated only slightly with thickness of external fat. When line and year effects were removed the same conclusion is reached from these data. Their findings as to the strength of the relationships between rate of gain and slaughter grade closely coincide with the results of this experiment.

A study was made by Cummings and Winters (*7*) to appraise certain factors related to carcass yields in swine. They reported that the best carcasses tended to come from those pigs making the fastest gains from birth to slaughter. They stated that there was a slight positive association between rate of growth and leanness of carcass in several of the breeds studied.

Donald (*8*) found breadth of eye muscle in swine to be independent of growth rate. In the present study when the correlations between gain and area of eye muscle were corrected for final weight there was no association between the two factors.

There was a stronger relationship between final weight and thickness of external fat over the ribeye than between gain and thickness of fat over the ribeye. When line and year effects were removed, the correlation coefficients for weaning weight and final weight with thickness of fat were very close to the same.

There was a positive association between both final weight and gain with dressing percentage, but when year and line effects were removed the correlation between dressing percentage and gain was not significant.

Length of body and length of leg had little or no relationship to gain when adjusted to a constant final weight.

It appears that some improvement in carcass quality should result under present methods of selection. When line and year effects were removed, final weight had a significant positive correlation with birth weight, weaning weight, gain, efficiency of gain, slaughter grade, carcass grade, area of eye muscle, thickness of fat over eye muscle, dressing percentage, and body measurements. None of the correlations could be considered high, but they were uniformly positive.

9

The length of time that selection should be directed toward increased final weight under a time-constant experiment is debatable. Perhaps this question can be answered best by a carcass appraisal at various stages of development.

Any associations between preslaughter and postslaughter evaluations are of particular importance in a selection program in beef cattle because of slow generation intervals, low reproductive rates, and cost of experimentation.

Efficiency

The importance of selection for efficiency has been stressed by a number of investigators. Winters and McMahon (27) pointed out the wide difference in the ability of steers to make economical gains. Black and Knapp (3) proposed a record of performance based on economy of gain from 500 to 900 pounds. Heritability estimates for efficiency of gain were made by Knapp and Nordskog (16) and revised by Knapp and Clark (17).

The positive association between efficiency and rate of gain was discussed in the preceding section. There was little association between efficiency and any of the other factors considered, either in the total correlations or in the correlations in which effects of the line, the year, or both, were removed. There was very little effect in correlations involving efficiency as evidenced by the close agreement between the total and within-line correlations.

The highest correlation coefficient in table 6 is a negative relationship between efficiency and dressing percentage. There are several possible explanations for the negative relationship, such as the tendency of the more efficient steers in this study to have slightly less external fat. Mason (22) pointed out that efficiency falls off as fattening increases. The necessity to theorize as to the explanation for this relationship shows the need for more detailed analyses in conjunction with complete growth records.

TABLE 6.—*Correlations between efficiency of gain and other factors studied* [1]

Factor	Total	Within lines	Within years	Within lines and years
Slaughter grade	0. 06	0. 06	0. 13*	0. 15*
Carcass grade	. 07	. 07	. 05	. 03
Area of eye muscle	−. 04	−. 03	. 04	. 07
Thickness of fat	−. 15*	−. 18*	−. 01	−. 03
Dressing percentage	−. 19*	−. 24*	−. 17*	−. 23*
Length of body	−. 16*	−. 16*	. 12*	. 06
Length of leg	−. 15*	−. 16*	. 02	. 06

[1] Asterisk indicates significant correlation at 1-percent level.

There was a low but significant correlation between efficiency and slaughter grade when year and line effects were removed. Black and Knapp (2) reported a stronger relationship than was obtained in this study, but they were working with a weight-constant population. Other correlations shown in table 6 were not statistically significant.

Relationships Between Grades and Subsequent Carcass Evaluations

Slaughter Grade

Correlation coefficients between slaughter grade and carcass evaluations are shown in table 7.

TABLE 7.—*Correlations between slaughter grade and other factors studied* [1]

Factor	Total	Within lines	Within years	Within lines and years
Carcass grade	0.54*	0.53*	0.51*	0.52*
Area of eye muscle	.16*	.18*	.25*	.29*
Thickness of fat	.22*	.24*	.39*	.50*
Dressing percentage	.19*	.24*	.31*	.38*
Length of body	.15*	.17*	.15*	.18*
Length of leg	.03	.05	.03	.06

[1] Asterisk indicates significant correlation at 1-percent level.

There was only a fair degree of association between slaughter and carcass grades, indicating that the slaughter grade had a relatively low predictive value for estimating carcass grade. Considering the fact that the slaughter and carcass grades were based upon the mean score of experienced committees, the agreement was not so close as expected. Cook et al. (6) reported a higher correlation coefficient between slaughter and carcass grades. They were working with steers killed at a constant weight, which may be responsible for the greater agreement between scores.

That carcass appraisal is more accurate than live animal appraisal, as shown by Lush et al. (20) is generally recognized. One of the most serious defects of grading live animals is that the grades do not indicate accurately the relative proportions of lean and fat in the carcass. When line and year effects were removed, the agreement between slaughter grade and area of eye muscle was only about one-half that of slaughter grade and thickness of fat over the eye muscle (table 7). The appearance of the animal at slaughter was influenced more by the amount of external fat over the ribeye than it was by the size of the ribeye itself.

Slaughter grade was significantly correlated with dressing percentage. Cook et al. (6) made the same observation from cattle killed at a constant weight.

There was no association between slaughter grade and length of leg. The correlation coefficient between slaughter grade and length of body was significant, although not high.

Carcass Grade

Carcass evaluation as a part of record of performance in beef cattle has been emphasized in the literature. Black and Knapp (2) reported that carcass grade was highly related to percent of edible carcass. Knapp and Clark (17) estimated the heritability of carcass

11

grade at 33 percent indicating the degree of inherent variation in steers of fairly similar quality.

The relative importance of carcass grading in evaluating the worth of the carcass has been studied by several investigators. McMeekan and Walker (23) included in their scoring system for beef carcasses an eye appraisal based upon marbling, color and texture of muscle, color and texture of fat and rib cover. Yeates (29) similarly suggested an eye appraisal as a supplement to body and ribeye measurements for complete carcass appraisal.

Correlation coefficients between carcass grades and other carcass evaluations are shown in table 8. When line and year effects were removed, carcass grade was significantly correlated with area of eye muscle, thickness of fat, and dressing percentage.

TABLE 8.—*Correlations between carcass grade and other factors studied* [1]

Factor	Total	Within lines	Within years	Within lines and years
Area of eye muscle	0. 08	0. 08	0. 21*	0. 23*
Thickness of fat	. 43*	. 43*	. 52*	. 54*
Dressing percentage	. 34*	. 36*	. 41*	. 45*
Length of body	. 04	. 05	. 08	. 09
Length of leg	—. 02	. 00	. 05	. 08

[1] Asterisk indicates significant correlation at 1-percent level.

Carcass grade and thickness of fat over the ribeye were correlated more highly than were carcass grade and area of ribeye. Several explanations are suggested. As area of ribeye and thickness of fat over the ribeye are both components of carcass grade it is possible that the thickness of fat was the greater determinant of carcass grade. Since the ultimate value of the carcass is enhanced more by a large eye muscle than by excess external fat, it is possible that thickness of external fat received too strong a consideration in the grading. However, another explanation for the higher correlation between grade and thickness of fat may be that thickness of fat is more closely associated with other factors that enter into a high carcass grade.

Length of body or length of leg was not significantly correlated with carcass grade. Cook et al. (6) in a study of live animal measurements reported that steers shorter in body and in height at withers and at the floor of the chest tended to grade slightly higher than more rangy steers. They studied data from steers killed at a constant weight.

Relationships Between Body Measurements and Certain Carcass Evaluations

Length of Body

Naumann (24) and Yeates (29) stressed the value of carcass measurements as an aid to improved appraisal of the effect of conformation on carcass value.

Length of body tends to be positively associated with height measurements but not with the circumference of foreflank or heart girth, according to Kohli et al. (*18*). They reported that they found wide individual variation in body dimensions.

The correlation coefficients for length of body and certain carcass evaluations are shown in table 9. The association between body length and length of leg, as measured in this experiment, was high. When adjusted to a constant final weight the correlation was much reduced, indicating that body proportions were not increasing proportionately.

TABLE 9.—*Correlations between length of body* [1] *and other factors studied* [2]

Factor	Simple correlations				Partial correlations [3]
	Total	Within lines	Within years	Within lines and years	Within lines and years
Area of eye muscle	0. 38*	0. 42*	0. 33*	0. 38*	0. 07
Thickness of fat	. 21*	. 22*	. 08	. 09	—. 13*
Dressing percentage	. 18*	. 25*	. 13*	. 21*	
Length of leg	. 77*	. 75*	. 75*	. 72*	. 37*

[1] A carcass measurement from the anterior edge of the first thoracic vertebra to the anterior edge of the aitch bone.
[2] Asterisk indicates significant correlation at 1-percent level.
[3] Computed with final weight held constant.

Length of body was significantly correlated with area of eye muscle although the correlation was not significant when adjusted to a constant final weight. The correlation between length of body and thickness of fat was not significant when line and year effects were removed, and there was a nonsignificant negative relationship when adjustment was made to a constant final weight.

The long-bodied steers in this experiment appeared to have carcasses as desirable as those of the short-bodied steers. The trend towards selection for short-coupled beef cattle could well be faulty considering the fact that most of the better cuts of beef are from the back.

Length of Leg

Total and partial correlations between length of leg and certain carcass evaluations are shown in table 10. In general they are similar to correlations between the same factors and length of body, particularly when line and year effects are removed.

Cook et al. (*6*), studying live animal measurements, stated that steers shorter in body and in height at withers and height at floor of chest tended to have a higher dressing percentage than more rangy steers. Again, the difference in experimental procedures undoubtedly influenced the relationships reported. The data collected by Cook et al. were from cattle killed at a constant weight. There was a positive

TABLE 10.—*Correlations between length of leg [1] and other factors studied [2]*

Factor	Simple correlations				Partial correlations [3]
	Total	Within lines	Within years	Within lines and years	Within lines and years
Area of eye muscle_____	0. 40*	0. 46*	0. 36*	0. 42*	0. 17*
Thickness of fat_____	. 13*	. 16*	. 02	. 04	—. 16*
Dressing percentage_____	. 12*	. 19*	. 08	. 16*	_____

[1] A carcass measurement from the anterior edge of the aitch bone to the highest point of the hock joint.
[2] Asterisk indicates significant correlation at 1-percent level.
[3] Computed with final weight held constant.

association between length of body and dressing percentage, also between length of leg and dressing percentage.

There was a significant positive association between length of leg and area of eye muscle, but not between length of leg and thickness of fat. When adjusted to a constant final weight the correlation with thickness of fat was negative. The same relationships existed between length of body and the above factors.

Relationships Between Certain Carcass Evaluations

One of the greatest obstacles to the detailed appraisal of a beef carcass is the expense involved. For this reason considerable attention has been centered on devising sampling methods by which a portion of the carcass can be used as an index of the value of the entire carcass. The 9th-10th-11th-rib cut seems to offer considerable promise in this regard.

The physical composition of the whole and edible portion of the 9th- 10th- 11th-rib cut is highly correlated with the physical composition of the empty body, carcass, and edible portion of the carcass according to Hopper et al. (*13*). The chemical composition of the 9th-10th-11th-rib cut was also found to be highly correlated with the chemical composition of the entire body.

Hankins et al. (*10*) found the muscle-bone ratio of the 9th-10th-11th-rib cut to be highly related to that of the entire carcass. Both beef and dual-purpose cattle were included in their data.

Hankins and Howe (*11*) reported a correlation coefficient of 0.85 between the separable lean of the 9th-10th-11th-rib cut and that of the entire carcass. Both steers and heifers were studied. For steers alone the correlation coefficient for these two factors was 0.90.

A further simplification in carcass appraisal has been recommended by a number of investigators and was used in this study. Under conditions in which more detailed study is not practicable, the determination of the area of the longissimus dorsi (eye muscle) between the 12th and 13th ribs has proven to be a very useful objective measurement of carcass value.

McMeekan and Walker (*23*) recommended a measure of the width of the eye muscle as a part of a beef carcass scoring system. Naumann (*24*) and Adams (*1*) both suggested measuring the area of the eye muscle as a means of carcass appraisal.

Area of Eye Muscle

Correlations between area of eye muscle and both thickness of fat over the eye muscle and dressing percentage are shown in table 11. There was a significant positive association between the area of eye muscle and dressing percentage which changed very little when line and year effects were removed. Area of eye muscle and thickness of fat over the eye muscle were not correlated. When final weight was held constant there was a slight negative relationship between them, although it was not statistically significant. From these results it would seem that selection for these two factors must be carried on independently.

Hankins and Burk (*9*) reported a high correlation between the thickness of external fat and the thickness of flesh of the carcass. They were working with a population covering a wide range in quality and finish, which may account for the different findings.

TABLE 11.—*Correlations between area of eye muscle and other factors studied* [1]

Factor	Simple correlations				Partial correlations [2]
	Total	Within lines	Within years	Within lines and years	Within lines and years
Thickness of fat over eye muscle_	0. 02	0. 01	0. 04	0. 01	−0. 09
Dressing percentage_____	. 39*	. 41*	. 34	. 36	_____

[1] Asterisk indicates significant correlation at 1-percent level.
[2] Computed with final weight held constant.

Thickness of External Fat

Lush et al. (*20*) reported that the percentage of fat in the wholesale rib cut was a very accurate indicator of the degree of fatness of the entire animal. This observation was later verified by Hopper et al. (*13*).

Hirzel (*12*) suggested that 20 and 12 millimeters were suitable depths for fat over the proximal and distal ends, respectively, of the eye muscle.

Correlations between thickness of fat over the eye muscle and both dressing percentage and shipping shrink are shown in table 12. When line and year effects were removed neither was found to be highly correlated with thickness of fat. Mason (*22*) concluded that dressing percentage was more influenced by the fatness of the animal than by any other factor. He reported a correlation coefficient between dressing percentage and amount of fatty tissue in the carcass that was somewhat higher than the coefficient obtained in this study. However, Lush (*19*) stated that dressing percentage was not a reliable indication of the proportion of fat within narrow limits of fatness.

TABLE 12.—*Correlations between thickness of external fat and other factors studied* [1]

Factor	Total	Within lines	Within years	Within lines and years
Dressing percentage	0. 41*	0. 39*	0. 31*	0. 25*
Shipping shrink	. 21*	. 18*	. 17*	. 12*

[1] Asterisk indicates significant correlation at 1-percent level.

There is conflicting evidence in regard to the relation between degree of fatness and tenderness. Mackintosh et al. (*21*) stated that there was a relationship between tenderness and amount of fat in the eye muscle. Branaman et al. (*4*) found no relationship between tenderness and fatness.

Color of Lean

Comfort (*5*) reported that dark colored meat was nearly equal in palatability to meat from brighter colored carcasses. He observed that color was influenced by amount of muscle hemoglobin, handling, and amount of exposure.

There were no strong relationships between color of lean and any of the production factors studied. However, when line differences for color were determined, they proved to be significantly different. Therefore it appears that color can be influenced by selection.

The meat trade strongly prefers a light colored lean. When carcasses are ribbed down before grading, the color of the eye muscle is taken into consideration in the grading and undoubtedly affected the correlation between carcass grade and color of lean, shown below.

TOTAL CORRELATIONS BETWEEN COLOR OF LEAN AND OTHER FACTORS STUDIED: [1]

Factor	Total	Factor	Total
Birth weight	0. 08	Slaughter grade	0. 17*
Weaning weight	. 13	Carcass grade	. 27*
Final weight	—. 03	Area of eye muscle	—. 03
Gain on test	—. 13*	Thickness of fat	—. 12*
Efficiency	—. 14*		

[1] Asterisk indicates significant correlation at 1-percent level.

Dressing Percentage

Contrary to prevailing opinion it is apparently not possible to estimate accurately individual dressing percentages prior to slaughter in groups of animals of fairly similar quality. The relationship between slaughter grade and dressing percentage was quite low. Carcass grade and dressing percentage were correlated somewhat more highly. These data are presented in tables 7 and 8.

SUMMARY AND CONCLUSIONS

Growth and carcass records collected on 635 Hereford steers from 1942–51 were studied to determine relationship between measures of production and carcass value at the United States Range Livestock Experiment Station, Miles City, Mont. These steers, which were uniform in type and quality, were marketed after a time-constant feeding period of 252 days each year. Relationships found in this study would undoubtedly differ from those in animals marketed at a constant weight or a constant finish.

Birth weight proved to be more highly correlated with subsequent growth than with carcass quality.

Calves that were heavier at birth grew more rapidly and their body proportions, i. e., length of body and length of leg, were larger at the end of the feeding period. They were slightly more efficient in feed conversion.

There was a tendency for the calves that were heavier at birth to have a slightly larger eye muscle, but less external fat. Indications are that in a weight-constant feeding trial these trends would be still more apparent.

Weaning weights varied widely by years, primarily owing to years of drouth. Weaning weight was less closely associated with subsequent rate of gain in the feedlot than was birth weight. It was negatively correlated with efficiency of gain, as would be expected. Weaning weight was the only production factor studied, exclusive of final weight, that had a significant positive correlation with the thickness of external fat over the ribeye. Indications are that there is no antagonism between heavy weaning weights and the ability to produce a desirable carcass.

Efficiency of gain was not significantly correlated with any of the measures of carcass quality. It was negatively correlated with dressing percentage.

There is evidence that inherent growth ability has a slight negative correlation with degree of external fattening. Area of eye muscle, on the other hand, was positively associated with gain. There was no correlation when adjustment was made to a constant final weight.

Slaughter grades proved to be a better estimate of thickness of external fat over the eye muscle than of area of eye muscle.

There is an indication that the amount of external fat was given too much weight, and the area of eye muscle not enough, in determining the carcass grade.

There was a surprising lack of correlation between area of eye muscle and thickness of fat over the eye muscle.

Length of body and length of leg were correlated approximately the same in regard to the other factors studied. Length of body was correlated slightly more with both rate of gain and final weight than was length of leg. However, neither were correlated with gain when final weight was held constant. Both measurements were related to area of eye muscle but not to thickness of fat over the eye muscle. When final weight was held constant, there was actually a negative correlation between both measurements and thickness of fat over the eye muscle. In this experiment the long-bodied steers appeared to have carcasses as desirable as those of the short-bodied steers.

17

Color of lean, as measured in the eye muscle, proved to be unrelated to most of the production factors studied. Differences in color of lean between lines were found.

Dressing percentages were not accurately estimated by the slaughter grades, nor did dressing percentage bear a strong relationship to the thickness of external fat.

In general, most of the desirable preslaughter and postslaughter characteristics were positively correlated. An exception may be the tendency of rapid growth to be negatively associated with the amount of external fat.

LITERATURE CITED

(1) ADAMS, C. H.
 1951. PHYSICAL, CHEMICAL AND ORGANOLEPTIC METHODS OF BEEF CARCASS
 EVALUATION. Ann. Rpt. Fourth Reciprocal Meat Conf.
(2) BLACK, W. H., and KNAPP, BRADFORD, Jr.
 1936. A METHOD OF MEASURING PERFORMANCE IN BEEF CATTLE. Amer. Soc.
 Anim. Prod. Proc. 72–77.
(3) ———
 1938. A COMPARISON OF SEVERAL METHODS OF MEASURING PERFORMANCE IN
 BEEF CATTLE. Amer. Soc. Anim. Prod. Proc. 103–107.
(4) BRANAMAN, G. A., HANKINS, O. G., and ALEXANDER, LUCY M.
 1936. THE RELATION OF DEGREE OF FINISH IN CATTLE TO PRODUCTION AND
 MEAT FACTORS. Amer. Soc. Anim. Prod. Proc. 295–300.
(5) COMFORT, J. E.
 1948. REVIEW OF CURRENT BEEF CARCASS RESEARCH. Ann. Rpt. First Reciprocal Meat Conf.
(6) COOK, A. C., KOHLI, M. L., and DAWSON, W. M.
 1951. RELATIONSHIP OF FIVE BODY MEASUREMENTS TO SLAUGHTER GRADE,
 CARCASS GRADE AND DRESSING PERCENTAGE IN MILKING SHORTHORN
 STEERS. Jour. Anim. Sci. 10: 386–394.
(7) CUMMINGS, J. N., and WINTERS, L. M.
 1951. A STUDY OF FACTORS RELATED TO CARCASS YIELDS IN SWINE. Minn.
 Agr. Expt. Sta. Tech. Bul. 195.
(8) DONALD, H. P.
 1940. GROWTH RATE AND CARCASS QUALITY IN BACON PIGS. Jour. Agr. Sci.
 30: 582–597.
(9) HANKINS, O. G., and BURK, L. B.
 1938. RELATIONSHIPS AMONG PRODUCTION AND GRADE FACTORS OF BEEF. U. S.
 Dept. Agr. Tech. Bul. 665.
(10) ——— KNAPP, BRADFORD, Jr., and PHILLIPS, RALPH W.
 1943. THE MUSCLE-BONE RATIO AS AN INDEX OF MERIT IN BEEF AND DUAL
 PURPOSE CATTLE. Jour. Anim. Sci. 2: 42–49.
(11) ——— and HOWE, PAUL E.
 1946. ESTIMATION OF THE COMPOSITION OF BEEF CARCASSES AND CUTS.
 U. S. Dept. Agr. Tech. Bul. 926.
(12) HIRZEL, R.
 1939. FACTORS AFFECTING QUALITY IN MUTTON AND BEEF WITH SPECIAL REF-
 ERENCE TO THE PROPORTIONS OF MUSCLE, FAT AND BONE. Onder-
 stepoort Jour. Vet. Sci. and Anim. Indus. 12: 379–550.
(13) HOPPER, T. H.
 1944. METHODS OF ESTIMATING THE PHYSICAL AND CHEMICAL COMPOSITION
 OF CATTLE. Jour. Agr. Res. 68: 239–268.
(14) KNAPP, BRADFORD, Jr., BAKER, A. L., QUESENBERRY, J. R., and CLARK, R. T.
 1941. RECORD OF PERFORMANCE IN HEREFORD CATTLE. Mont. Agr. Expt. Sta.
 Bul. 397.
(15) ——— PHILLIPS, RALPH W., BLACK, W. H., and CLARK, R. T.
 1942. LENGTH OF FEEDING PERIOD AND NUMBER OF ANIMALS REQUIRED TO
 MEASURE ECONOMY OF GAIN IN PROGENY TESTS OF BEEF BULLS. Jour.
 Anim. Sci. 1: 285–292.
(16) ——— and NORDSKOG, A. W.
 1946. HERITABILITY OF GROWTH AND EFFICIENCY IN BEEF CATTLE. Jour.
 Anim. Sci. 5: 62–70.
(17) ——— and CLARK, R. T.
 1950. REVISED ESTIMATES OF HERITABILITY OF ECONOMIC CHARACTERISTICS OF
 BEEF CATTLE. Jour. Anim. Sci. 9: 582–587.
(18) KOHLI, M. L., COOK, A. C., and DAWSON, W. M.
 1951. RELATIONS BETWEEN SOME BODY MEASUREMENTS AND CERTAIN PER-
 FORMANCE CHARACTERS IN MILKING SHORTHORN STEERS. Jour.
 Anim. Sci. 10: 352–364.
(19) LUSH, J. L.
 1926. PRACTICAL METHODS OF ESTIMATING PROPORTIONS OF FAT AND BONE IN
 CATTLE SLAUGHTERED IN COMMERCIAL PACKING PLANTS. Jour. Agr.
 Res. 32: 727–755.
(20) ——— BLACK W. H., and SEMPLE, A. T.
 1929. THE USE OF DRESSED BEEF APPRAISALS IN MEASURING THE MARKET
 DESIRABILITY OF BEEF CATTLE. Jour. Agr. Res. 39: 147–162.

(21) MACKINTOSH, D. L., HULL, J. L., and VAIL, G. E.
   1936. SOME OBSERVATIONS PERTAINING TO PROTEIN TENDERNESS OF MEAT.
      Amer. Soc. Anim. Prod. Proc. 285–289.
(22) MASON, I. L.
   1951. PERFORMANCE RECORDING IN BEEF CATTLE. Anim. Breeding Abs. 19:
      1–24.
(23) McMEEKAN, C. P., and WALKER, D. E.
   1950. JUDGING BEEF CARCASSES BY MEASUREMENT. Past. Rev. 60: 795–797.
(24) NAUMANN, H. D.
   1951. A RECOMMENDED PROCEDURE FOR MEASURING AND GRADING BEEF FOR
      CARCASS EVALUATION. Ann. Rpt. Fourth Reciprocal Meat Conf.
(25) PONTECORVO, GUIDO.
   1936. LA VALUTAZOINE DEI TORI NELLE RASSE DA CARNE A CARNELAVORO
      MEDIANTE LO STUDIO DELL A DISCENDENZE. Ministero Dell' Agr.
      E. Dalle For. Anno XIII 6: 7–26.
(26) STANLEY, E. B., and McCALL, RALPH.
   1945. A STUDY OF PERFORMANCE IN HEREFORD CATTLE. Ariz. Agr. Expt.
      Sta. Tech. Bul. 109.
(27) WINTERS, L. M., and McMAHON, HARRY.
   1933. EFFICIENCY VARIATIONS IN STEERS. Minn. Tech. Bul. 94.
(28) WOODWARD, R. R., CLARK, R. T., and CUMMINGS, J. N.
   1942. STUDIES ON LARGE AND SMALL TYPE HEREFORD CATTLE. Mont. Agr.
      Expt. Sta. Bul. 401.
(29) YEATES, N. T. M.
   1952. THE QUANTITATIVE DEFINITION OF CATTLE CARCASSES. Austral. Jour.
      Agr. Res. 33: 68–94.

TABLE 1.—*Preslaughter records by years*

Year	1942	1943	1944	1945	1946	1947	1948	1949	1950	1951
Number of steers	62	45	46	53	63	98	85	88	50	45
Birth weight (pounds)	81.9	83.4	79.2	80.2	79.9	82.8	83.9	81.7	74.9	74.8
Weaning weight (pounds)	415.6	401.7	391.0	373.6	373.3	402.3	419.2	433.8	319.2	393.8
Final weight (pounds)	919.3	¹887.9	¹848.2	866.8	864.8	958.9	945.7	966.2	916.9	1,017.2
Gain on test (pounds)	504.5	¹493.0	¹462.0	494.4	483.8	574.2	548.2	549.5	593.0	619.2
Efficiency of gain	19.39	18.40	18.17	19.94	19.25	18.78	18.00	18.01	19.47	18.83
Sale weight (pounds)	889.0	899.3	827.8	827.2	818.2	927.6	918.8	922.5	883.9	956.9
Shrinkage (pounds)	30.4	24.0	22.9	39.7	46.6	31.5	26.9	43.7	33.0	60.3
Slaughter grade	15.7	13.1	15.0	14.4	14.3	14.0	14.6	15.3	17.2	14.8

¹ Adjusted to 252 days.

TABLE 2.—*Postslaughter records by years*

Year	1942	1943	1944	1945	1946	1947	1948	1949	1950	1951
Number of steers	62	45	46	53	63	98	85	88	50	45
Carcass weight (pounds)	537.0	545.0	493.7	491.1	492.0	558.8	555.6	557.2	544.0	605.6
Dressing percentage (percent)	58.37	59.01	58.00	56.65	56.88	58.26	58.72	57.64	59.31	59.51
Carcass grade	16.2	13.7	15.2	14.5	14.8	12.4	15.0	17.1	16.2	13.1
Color of lean	4.5	3.9	2.8	5.5	5.1	4.2	5.5	4.4	5.8	5.7
Area of eye muscle (square inch)	12.36	11.45	9.42	9.48	10.30	10.70	11.57	10.74	10.63	10.60
Thickness of fat (millimeter)	9.4	11.7	11.8	11.5	10.8	15.8	17.4	15.6	17.5	21.0
Length of body (centimeter)	-	-	-	114.1	114.8	116.0	116.6	117.2	114.6	117.6
Length of leg (centimeter)	-	-	-	71.4	72.1	72.3	72.9	73.3	71.9	73.0

CPSIA information can be obtained
at www.ICGtesting.com
Printed in the USA
BVHW040908141218
535629BV00023B/1333/P